Let's Look at Lizards

MONITOR LIZARDS

By Seth Lynch

Gareth Stevens
PUBLISHING

Please visit our website, www.garethstevens.com. For a free color catalog of all our high-quality books, call toll free 1-800-542-2595 or fax 1-877-542-2596.

Library of Congress Cataloging-in-Publication Data
Names: Lynch, Seth, author.
Title: Monitor lizards / Seth Lynch.
Description: Buffalo, New York : Gareth Stevens Publishing, [2025] | Series: Let's look at lizards! | Includes bibliographical references and index.
Identifiers: LCCN 2023034873 | ISBN 9781538292792 (library binding) | ISBN 9781538292785 (paperback) | ISBN 9781538292808 (ebook)
Subjects: LCSH: Monitor lizards–Juvenile literature.
Classification: LCC QL666.L29 L96 2025 | DDC 597.95/96–dc23/eng/20230728
LC record available at https://lccn.loc.gov/2023034873

First Edition

Published in 2025 by
Gareth Stevens Publishing
2544 Clinton Street
Buffalo, NY 14224

Editor: Brianna Propis
Designer: Leslie Taylor

Photo credits: Cover elena_photo_soul/Shutterstock.com; p. 5 Martin Pelanek/Shutterstock.com; p. 7 Ken Griffiths/Shutterstock.com; p. 9 Carolyn Smith1/Shutterstock.com; p. 11 (top) Imogen Warren/Shutterstock.com, (lower left) reptiles4all/Shutterstock.com, (lower right) TashaBubo/Shutterstock.com; p. 13 Greens and Blues/ Shutterstock.com; p. 15 CKChong/Shutterstock.com; p. 17 Sergey Uryadnikov/Shutterstock.com; p. 19 Sergey Uryadnikov/Shutterstock.com; p. 21 Suphakornx/Shutterstock.com; p. 23 Dr David Sing/Shutterstock.com.

Printed in the United States of America

CPSIA compliance information: Batch #CSGS25: For further information contact Gareth Stevens, New York, New York at 1-800-542-2595.

Contents

Monitor lizards
are very big.
They are smart.

They have scales.
They have long necks.
They have claws.

They can see very well.
They do not blink.

They can be brown
or gray.
They can be yellow
or blue.
Their colors help
them hide.

They live in forests.
They live in swamps.

They eat fish and birds.
They eat bugs and eggs.

They can stand on
two legs.
This scares animals away.

They are many kinds.
Some are bigger
than others!

Mothers lay eggs. They hide them in tree trunks.

Babies stay in trees.
It keeps them safe.

Words to Know

claws

scales

swamp

Index